YOUR KNOWLEDGE HAS VALUE

Frank Kostrzewa

Famous mathematicians and their problems

GRIN Publishing

Bibliographic information published by the German National Library:

The German National Library lists this publication in the National Bibliography; detailed bibliographic data are available on the Internet at http://dnb.dnb.de .

Imprint:

Copyright © 2014 GRIN Verlag GmbH
Print and binding: Books on Demand GmbH, Norderstedt Germany
ISBN: 978-3-656-82050-5

This book at GRIN:

http://www.grin.com/en/e-book/282774/famous-mathematicians-and-their-problems

GRIN - Your knowledge has value

Since its foundation in 1998, GRIN has specialized in publishing academic texts by students, college teachers and other academics as e-book and printed book. The website www.grin.com is an ideal platform for presenting term papers, final papers, scientific essays, dissertations and specialist books.

Visit us on the internet:

http://www.grin.com/

http://www.facebook.com/grincom

http://www.twitter.com/grin_com

Famous mathematicians and their problems

I.   The Ishango bone:

The Ishango bone which was found at Lake Edward in Zaire, close the Ugandan border, in 1960 is about 20.000 years old.

Not only the age of the bone is remarkable but also the numbers on this bone illustrated as notches.

At one place we see the numbers 3-6, 4-8 and 10-5. These are obviously the representations of the doubling and the bisection of numbers.
Another place on the bone is even more remarkable. We find the numbers 11-13-17-19, all the prime numbers between 10 and 20.

So far nobody really knows what these prime numbers were used for 20.000 years ago.

II.   The calculating system of the Babylonians

The first sensible number was used by the Babylonians 4000 years ago.

The system was developed in Mesopotamia which is Iraq nowadays.

The Babylonians had something similar to our decimal system, but for them not 10 was the decisive number, but 60.

This means they had a positional notation system to the base of 60.

Some relics of the Babylonian number system are still relevant today. Thus one hour consists of 60 minutes and 60 minutes consist of 60 seconds, so that an hour consists of 60 x 60 = 3600 seconds.

A circle consists of 360 (6 x 60) angular degrees.

All this has survived for 4000 years.

A number system to the base of 60 has the following consequences:

In this system the numbers 1 to 59 are used.

The value of a number is dependent on its position.

The number 5 in final position (unit position) also has a value of 5.

The last but one position is not, as in our decimal system, the number 10, but the number 60.

A 2 in this position leads to a value of 120.

The third last position in the decimal system is the 100 position.

In the Babylonian number system to the base of 60 the third last position is 3600.

A 3 in this position would have a value of 3 x 3600 = 10800.

With this system the Babylonians could add, subtract, multiply and divide. There was just one main thing missing: cipher

III. The unsolvable mathematical problems of the ancient world:

The doubling of the cube:

A cube has the (side) length of 1 and thus also the volume of 1.
The task is to construct a cube with double the volume.
This cube would have to have edges, the lengths of which are the third root of 2.
The question is whether it is possible to construct this length (third root of 2) by only using compass and ruler.

The angle trisection:

Is it possible to construct for each random angle an angle which is exactly one third of the original one?
For instance in order to trisect the angle of 45 degrees we would have to construct an angle with 15 degrees. This could relatively easily be done because we would only have to bisect the angle of 60 degrees twice. But does it work for all angles?

Squaring the circle:

The problem of whether it is possible to square the circle has been around for more than 2000 years.

Euclid mentions this problem in his book "Elements" and he postulates that only ruler and compass should be allowed to solve the problem.

In the year 1882 the mathematician Ferdinand Lindemann claimed that he had found a solution to this problem, though only a negative one. In the sense that the problem of squaring the circle is not solvable.

IV. The five Platonic solids:

(1) Tetrahedron:

number of surfaces: 4

number of angles: 4

ancient element: fire

(2) Hexahedron:

number of surfaces: 6

number of angles: 8

ancient element: earth

(3) Octahedron:

number of surfaces: 8

number of angles: 6

ancient element: air

(4) Icosahedron:

number of surfaces: 20

number of angles: 12

ancient element: water

(5) Dodecahedron:

number of surfaces: 12

number of angles: 20

ancient element: universe

V.   The Pythagoreans: Amicable numbers

The Pythagoreans were fascinated by amicable numbers like 220 and 284. In this pair of numbers 220 can be divided by 1, 2, 4, 5, 10, 11, 20, 22, 44, 55 and 110, leading to a sum of 284. 284 can be divided by 1, 2, 4, 71 and 142 leading to a sum of 220.
Until the year 1747 the Swiss mathematician and physicist, Leonhard Euler, reported that there is a limited number of only 30 amicable pairs. Today with the help of modern computers mathematicians have found more than 11 million amicable pairs, but in only 5001 pairs both numbers are smaller than $3,06 \times 10^{11}$.
In the year 850 the Arabian astronomer and mathematician Thabit ibn Qurra presented a formula according to which amicable numbers can be calculated:
If $p = 3 \times 2^{(n-1)} - 1$, and $q = 3 \times 2^{(n)} - 1$, and $r = 9 \times 2^{(2n-1)} - 1$ for a number $n > 1$

→ It follows that if p, q and r are prime numbers:

The numbers $2^{(n)} pq$ and $2^{(n)} r$ are amicable numbers.

VI.   Pythagorean triples:

In the sequence of uneven whole numbers 1,3,5,7,9,11, the sum of succeeding elements starting from 1 results in a square number:

3

$1 = 1\char`\^2$

$1 + 3 = 4 = 2\char`\^2$

$1 + 3 + 5 = 9 = 3\char`\^2$

$1 + 3 + 5 + 7 = 16 = 4\char`\^2$

$1 + 3 + 5 + 7 + 9 = 25 = 5\char`\^2$

$1 + 3 + 5 + 7 + 9 + 11 = 36 = 6\char`\^2$

$1 + 3 + 5 + 7 + 9 + 11 + 13 = 49 = 7\char`\^2$

Pythagorean triples:

$3\char`\^2 + 4\char`\^2 = 5\char`\^2$

$5\char`\^2 + 12\char`\^2 = 13\char`\^2$

VII.   Euclid of Alexandria on perfect numbers:

Is $1 + 2 + 2\char`\^2 + 2\char`\^3 + ... + 2\char`\^$ (n-1) a prime number,
then $(1 + 2 + 2\char`\^2 + 2\char`\^{}+3 + ... + 2\char`\^$ (n-1)) x $2\char`\^$ (n-1) is a perfect number.

VIII.   Euclid's Golden Ratio:

In the 6th book of the **"Elements" by Euclid**, we find a definition of a particular type of partition of a line segment in two uneven parts. According to Euclid a line segment AB can be divided by an interior point C.
If this is the case, then the line segment AB is equicontinuously divided if the quotient AC/CB is equal to the quotient AB/AC.
Since the 19th century this relation has been called the "golden ratio" or "golden section".
The value of the golden ratio is $(1 + \sqrt{5})/2 = 1,6180339887 \dots$

IX.   The Sand Calculator:

Apart from his ideas on many mathematical problems, Archimedes of Syracuse is also known for calculating with enormously big numbers. In his treatise "The Sand Calculator" Archimedes calculated how many grains of sand would be needed to fill the universe. Archimedes calculated that approximately 8 x $10\char`\^63$ grains of sand would be needed to fill the universe.

X.   The Bakhshali-Manuscript:

The Bakhshali-Manuscript is a remarkable collection of mathematical scripts written in the 3rd century A.D. It describes methods and rules to solve mathematical problems and presents exercises for the user. One of these exercises reads as follows: We have a group of altogether twenty men, women and children. Together they earn twenty coins. Each man earns three coins, each woman one and a half and each child half a coin. Then the author of the exercise formulates his question: How many men, women and children do we find in this group? The author then presents two formulas as a hint:

m + f + k (men + women + children) = 20 and
3m + (3/2) f + (1/2) k = 20.

There is only one solution for this problem: We have two men, five women and thirteen children in the group.

XI.   The Indians and the number cipher:

Since when has the number cypher been in use?
The number cypher was definitely invented in India. The first well-documented Indian cypher can be found in a Vishnu temple in Gwalior, 400 kilometers south of Delhi. On a stone tablet marked with the year 876 the number cypher is used to illustrate the numbers 50 and 270. The Arabs adopted the number cypher from the Indians and with the expansion of the Islam the cypher also came to Europe.
In the year 1202 at the latest the number cypher had arrived in Western Europe. In this year the book "Liber abaci" by the arithmetician Leonardo of Pisa was published. Leonardo of Pisa later on became famous under the name of Fibonacci.
Fibonacci wrote in his book: The nine Indian numbers are 9 8 7 6 5 4 3 2 1. With these nine numbers and the number 0, which the Arabs call "Zephirum" any number whatsoever can be written.

XII.   Indian-Arabian number system:

We all can be glad that we use Indian-Arabian numbers nowadays which were mainly developed in the 6th and 7th century.

We all remember the Roman numbers:

I: 1

V: 5

X: 10

L: 50

C: 100

D: 500

M: 1000

Thus 1492 would be MCDXCII.

With the Roman symbols it was very difficult to carry out even simple additions and subtractions.

Even the Greeks who were masters of geometry had a highly complicated number system:

I: 1

P: (pente = five)

D: (deka = ten)

H: (hekatón = one hundred)

Thus 428 in the Greek system would be HHHHDDPIII.

XIII. Ibn Al-Haitham (965 - 1039) on cubic numbers:

Al-Haitham multiplied the sum of the first k cubic numbers with the next number:

$(1^3 + 2^3 + ... + k^3) (k + 1)$

$= 1^3 (1 + 1 + ... + 1)$

$+ 2^3 (2 + 1 + ... + 1)$

$+ ...$

$+ (k - 1)^3 ((k - 1) + 1 + 1)$

$+ K^3 (k + 1)$

$= (1^4 + 2^4 + ... + k^4)$

$+ 1^3 + (1^3 + 2^3)$

$+ (1^3 + 2^3 + 3^3)$

$+ ...$

Thus he combined the sum of numbers raised to the power of four with different sums of numbers raised to the power of three.

XIV. Abu Ali Al-Husain Ibn Sina (980 - 1037) on perfect numbers:

A natural number is a perfect number if the number itself is the sum of its proper divisors:

$6 = 1 + 2 + 3$
$28 = 1 + 2 + 4 + 7 + 14$

Perfect numbers are also the sum of immediately consecutive numbers, e.g.:

$6 = 1 + 2 + 3$

$28 = 1 + 2 + 3 + 4 + 5 + 6 + 7$

The sum of the reciprocals of all proper divisors of a perfect number is always 2:

$1/1 + 1/2 + 1/3 + 1/6 = 2$

$1/1 + 1/2 + 1/4 + 1/7 + 1/14 + 1/28 = 2$

XV. Cardano's „Ars Magna":

The Italian mathematician and physicist Gerolamo Cardano (1501-1576) is still today well-known for his work on algebra by the title of "Artis magnae, sive de regulis algebraicis" (Of the Great Art – or the algebraic rules). The short form of the title nowadays used is "Ars magna".
In "Ars magna" Cardono wrote about the existence of numbers which we would nowadays call imaginary numbers based on the square root of -1.
Cardono presents a calculation with complex numbers when he writes:
If you multiply $5 + \sqrt{-15}$ with $5 - \sqrt{-15}$, you will get $25 - (-15) = 40$.
During the time of the Inquisition Cardano spent several weeks in prison because he had cast Jesus Christ's horoscope.
He also calculated his own day of death, but when he noticed that he was still in good health on the calculated day, he committed suicide.

XVI. Blaise Pascal's probability calculation

Blaise Pascal:

Blaise Pascal, like other philosophers and mathematicians, was interested in the possibility of predicting future events.

Pascal was born in Clermont and lived from 1623 to 1662.

His central work is the Pensées, which was first published under the title of Pensées sur la religion et autres sujets in 1670, eight years after Pascal's death.

In the second chapter of this book Pascal deals with the dialectic of gambits, later on also called "Pascal's Gambit".

Already at the age of twelve Pascal had published a much quoted work on conic sections.

Further much quoted works were:
Potestatum numericarum summa
De numeris multiplicibus

Pascal's first interest in probability calculation arose when a close friend, Chevalier de Marre, who was addicted to gambling, asked Pascal to help him improve his chances in roulette.

Pascal reflected about this problem in much depth and the famous Pascal's triangle can be regarded as a consequence and result of these considerations.

But since Pascal was also a theologist he was also interests in finding a proof for the existence of God.

Pascal claimed that like the infinity of numbers, humans are not able to conceive the infinity of God.

It is not possible to find conclusive evidence for the existence of God. On the other hand it is also not possible to find conclusive evidence for his non-existence.

Since the non-existence of God cannot be conclusively proved, it would be sensible to believe in God without demanding further evidence for his existence.

Looked at it objectively the chances for the existence of God are 50:50.

If we decide to believe in God we have the following constellation:

I decide to believe in God and can potentially win (i.e. God exists).

If God really exists I would gain infinitely much (by my belief).

If God does not exist I would have lost relatively little (by my belief).

Coming to a conclusion we could say that with my decision for a belief in God I have the chances of a potentially high benefit (if God exists).

On the other hand my damage is kept in bounds if God does not exist. If I decide against my belief in God we have the following constellation:

I decide against God (my maximum benefit would be to be proved right)

If God does not exist I have not lost anything.

If God however exists I have lost infinitely much.

Coming to a conclusion we could claim that the chances of winning by not believing in God are infinitely small but the dangers of losing infinitely much by not believing in God are infinitely high.

So, Pascal concluded, it would only be reasonable to prophylactically believe in God.

Pascal's Probability Caculation:

If you throw dice, the probability of throwing a six with one cast of dice is 1/6 or, in other words, 16, 7%.

This means that the probability of not throwing a six is 83, 3%.

Since two consecutive casts of dice are independent of each other, the probability of not having a six in two consecutive casts is 5/6 x 5/6.

The probability of not having a six in ten consecutive casts of dice would then be (5/6) ^10 = 0, 16.

This means that in 16% of all cases you would not even get a six in ten consecutive casts of dice but then again in 84% of all cases you would at least get one six in ten consecutive casts.

Mathematicians who specialize in probability calculation have found out that the relative frequencies stabilize as numbers (in this case casts of dice) increase.

With increasing numbers (let's assume you cast your dice one million times) the variance of results decreases leading to stable percentages.

Mathematicians call this regularity the law of large numbers.

XVII.   John Napier (1550 - 1617) Logarithm:

John Napier can be regarded as the inventor of the logarithm. The equation $e^{\wedge}$ (ln N), though also attributed to Leonhard Euler influenced mathematics enormously. Before Napier introduced the term "logarithm" he used the notion "artificial numbers". When he applied the term "logarithm" he used it in the sense of "ratio" or "relational number" since the logarithms are defined by their numerical relationship.

Thus a: b = c:d $\leftrightarrow$ log (a) - log (b) = log (c) - log (d)

XVIII.   Kepler on Fibonacci numbers

The astronomer Johannes Kepler (1571-1630) was the first scientist who noticed that Fibonacci numbers occur in nature in many surprising ways. He claimed that if we count the number of petals of different flowers we often come across Fibonacci numbers to such an extent which can no longer be regarded as pure coincidence.

The Iris for example normally has 3 petals, the primrose, the buttercup, the dog rose or Alpine rose, the larkspur and the aquilegia have 5 petals each. Asters, black-eyed Susans and coffeeweed have 21 petals each. Daisies have 13, 21 or 34 petals.

Plantains and chrysanthemums have 34 petals each.
New York Asters have 55 or 89 petals.

What they all have in common is that the number of their petals are Fibonacci-Numbers.

XIX. Kepler on the relationship between the Golden Ration and the Fibonacci numbers

Golden Ratio and Fibonacci-numbers:

Two thousand years after Euclid's death the German astronomer Johannes Kepler discovered that the Golden Ratio (1,618 ...) is connected to a sequence of numbers called Fibonacci-numbers.
If you take the first numbers of the Fibonacci sequence you have:

1, 1, 2, 3, 5, 8, 13, 21, 34, 55, 89, 144, 233, 377, ...

From the third element of the sequence onwards each number is the sum of its two predecessors:

$2 = 1 + 1$

$3 = 1 + 2$

$5 = 2 + 3$

If you now divide a number by its immediate predecessor you will always get a value which is close to the Golden Ratio:

$144 : 89 = 1,617978$

$233 : 144 = 1,618056$

$377 : 233 = 1,618026$

XX. Bernoulli's inequation

Jakob Bernoulli (1655 - 1705):

Most students connect the name of Jakob Bernoulli with the following inequation named after him:

For $x > -1$ $(x \neq 0) \rightarrow (1 + x)^n > 1 + nx$

But he also dealt with harmonic analysis of the following kind:

$1 + 1/2 + 1/3 + 1/4 + ...$

These sequences grow continuously.

The reciprocals of square numbers, however, converge towards a limit:

$1 + 1/4 + 1/9 + 1/16 + ... < 2$

XXI. Christian Goldbach's conjecture

Christian Goldbach:

In a letter dating from 7th June 1742 the Prussian amateur mathematician Christian Goldbach claimed that all whole even numbers greater than 2 could be written as the sum of two prime numbers (Goldbach conjecture). Although the British mathematician G.H. Hardy claimed that the conjecture was so simple that any idiot could have made this assumption the proof of the Goldbach conjecture has so far not been completed.

The validity of the assumption can be proved easily for the first even numbers greater than 2:

4 = 2 + 2 (correct)

6 = 3 + 3 (correct)

8 = 3 + 5 (correct)

10 = 3 + 7 or 5 + 5 (correct)

12 = 5 + 7 (correct)

14 = 3 + 11 or 7 + 7 (correct)

16 = 5 + 11 or 3 + 13 (correct)

XXII. The Goldbach Conjecture:

Weak Goldbach Conjecture:

In 1742 the Prussian historian and mathematician Christian Goldbach assumed that all whole numbers > 5 could be written as the sum of three prime numbers: Thus 21 would be the sum of 11 + 7 + 3

Strong Goldbach Conjecture:

In a new conjecture, nowadays called "strong Goldbach conjecture", the Swiss mathematician Leonhard Euler claimed that all positive even numbers > 2 could be written as the sum of two prime numbers.

So far the validity of the strong Goldbach conjecture could not be proved, but with the help of a computer programme Tomas Oliviera e Silva could show that it is at least valid for all numbers up to 12×10^{17}.

XXIII. The Euler number e:

The Euler number e can be determined as approximately 2, 71828. This number is also the limit value of $(1 + 1/n)$ with the nth power for n being infinitesimal.

Euler was the first mathematician who constantly used the number e from 1727 onwards. In 1737 Euler showed that e is irrational, i.e. it cannot be expressed as the relationship of two integral numbers.

In 1748 Euler found 18 fractional digits of e. Today more than 1.000.000.000.000 of these fractional digits are known.

E can also be expressed as $e^{\wedge}(i\pi) + 1 = 0$ for i being the square root of (-1).

XXIV.   The Stirling Formula (here also **Euler** number e is needed):

For all natural numbers n, n! is the product of natural numbers, e.g.:
4! = 1 x 2 x 3 x 4 = 24

The notation n! was introduced by the French mathematician Christian Kramp in 1808.

The value of a factorial number can be very high:
70! > $10^{\wedge}100$
25.206! > $10^{\wedge}100000$

For calculating n! Stirling introduced the following formula:
n! = $\sqrt{}$ (2π) x $e^{\wedge}$ (-n) x $n^{\wedge}$ (n+1/2), assuming a value of 2, 71828 for e and 3, 14159 for π.

XXV.   Gauß (and others) on Prime Numbers

How many prime numbers are there?

We know that a prime number is a natural number > 1, which can only be divided by 1 and itself.

The first prime numbers are: 2, 3, 5, 7, 11, 13, 17, and 19.

How are the prime numbers distributed in the range of numbers and what is their density in their respective range of numbers?

It is known that when the natural numbers increase, the amount and density of prime numbers decreases.

But if you look at certain intervals of numbers there are still more prime numbers than square numbers.

For the distribution of prime numbers the following can be stated:

Between 1 and 100 there are altogether 25 prime numbers.

Between 1000 and 1100 there are altogether 16 prime numbers.

Between 10000 and 10100 there are altogether 11 prime numbers.

Between 100 000 and 100 100 there are only 6 prime numbers.

Carl Friedrich Gauß and later Hadamard and de la Vallée Poussin claimed and proved that there is an enormous amount of prime numbers if we look at the range of numbers up to $10^{\wedge}100$.

The Prime Number Theorem claims that in the range of numbers up to 10^100 1% of those numbers are prime numbers.

This means that the likelihood of a 100-digit-number being a prime number is 1% of 10^100, which still is 10^98 or 4,3 x 10^97.

XXVI.   Gauß' Prime Number Conjecture

The Prime Number Conjecture:

Π (n) is the set of prime numbers ≤ a given number n
In 1792 the 15-year-old Johann Carl Friedrich Gauß focussed on prime numbers and formulated the conjecture that π (n) ≈ n/ln (n) with ln (n) being the natural logarithm of n.

Gauß followed that the n-th prime number should be ≈ nln (n). Later on Gauß corrected his approximation into π (n) ≈ Li (n) with Li (n) being the integral from 2 to n over dx/ln(x).

In 1914 Littlewood claimed that the difference of Li (n) - π (n) leads to a change of number sings with an increasing n.

In 1933 Stanley Skewes showed that the first change of signs happens with the number 10^10^10^34.

This number is called the Skewes number from that time on.

XXVII.   Lagrange Theorems:

The 4-Squares-Theorem:

Every natural number can be described as the sum of a maximum of four square numbers.

Prime-Number-Theorem:

A natural number n is a prime number if (n-1)! can be divided by n.

Theorem for the equation $Dx^2 + 1 = y^2$

In the set of whole numbers the equation $Dx^2 + 1 = y^2$ is solvable if D is a natural non-squared number.

XXVIII.   Joseph-Louis Lagrange: Binominal Theorem

The French mathematician Joseph-Louis Lagrange was the eldest of eleven children and was raised in Torino (Italy). His father wanted him to become a lawyer but at the age of 17 Joseph-Louis decided to dedicate his life to mathematics.
At the age of 18 he wrote a letter in Latin to the German mathematician Leonhard Euler who lived in Berlin and explained that he had found an analogy between the binomial theorem and

the higher derivatives of a product of functions.

If f and g are functions:

$(fg)' = f'g + fg'$

$(fg)'' = f''g + 2 f'g' + fg''$

$(fg)''' = f''g + 3 f'g + 3 fg'' + fg'''$

Generally speaking:

$(f \times g)^\wedge (n) = (n/0) \times f^\wedge (n) \times g^\wedge (0) + (n/1) \times f^\wedge (n-1) \times g^\wedge (1) + (n/2) \times f^\wedge (n-2) \times g^\wedge (1) + ... + (n/n) \times f^\wedge (0) \times g^\wedge (n)$

With $f^\wedge(n)$ being the nth derivative and $f^\wedge(0)$ being the function itself.

Lagrange, however, was very disappointed when Euler answered him that he himself and Johann Bernoulli had already found this regularity.

XXIX.   Lagrange Theorems: The 4-Squares-Theorem:

The 4-Squares-Theorem:

Every natural number can be described as the sum of a maximum of four square numbers.

Prime-Number-Theorem:

A natural number n is a prime number if (n-1)! can be divided by n.

Theorem for the equation $Dx^2 + 1 = y^2$

In the set of whole numbers the equation $Dx^2 + 1 = y^2$ is solvable if D is a natural non-squared number.

XXX.   Jurij Vega (I): Logarithms

Jurij Vega (also Latin: Georgius Vecha) has become famous for his work on logarithms. In 1783 he published his first table of logarithms containing seven-digit-logarithms to the base of 10. His work, comprising more than 400 pages, contained the logarithms of the natural numbers from 1 to 101 000 as well as the logarithms of the trigonometric functions for angular degrees between 0 and 45 with a step range of 10 arc seconds. The preface of his book contained altogether forty pages and was written in Latin and German. In 1794 Vega published his third table of logarithms called "Thesaurus logarithmorum completes", containing ten-digit-logarithms needed for special calculations in astronomy. On the basis of Vega's work James Gregory and Brook Taylor developed a method of describing functions on the basis of their derivations (Taylor series). For natural logarithms to the base of e the following series development can be described:

$Ln (1 + x) = x - (x^2/2) + (x^3/3) - (x^4/4) + (x^5/5) \ldots$

On the basis of the relationship:

$$\ln(1-x) = -x - (x^2/2) - (x^3/3) - (x^4/4) - \ldots$$

We get the quickly converging progression:

$$\ln((1+x)/(1-x)) = 2((x + (x^3/3) + (x^5/5) + \ldots)$$

XXXI: Jurij Vega (II):

If we want to calculate ln (2) we first have to determine the corresponding value of x:
$(1+x)/(1-x) = 2 \leftrightarrow x = 1/3$

Then we can calculate:
$\ln(2) = 2(1/3 + (1/3 \times 3^3) + (1/5 \times 3^5) + (1/7 \times 3^7) + \ldots)$
With the first ten summands we get an exact value for the first ten digits.

The relationship between the decadic and the natural logarithm is the following one:
$\lg(x) = \log_{\sqrt{10}}(x) = \log_{\sqrt{e}}(x)/\log_{\sqrt{e}}(10) = \ln(x)/\ln(10)$.

In order to determine the decadic logarithm of a number x the natural logarithm of x has to be determined first. In a second step the natural logarithm of x is then multiplied with the inverse value of the natural logarithm of 10:

$1/\ln(10) = 1/\ln(2) + \ln(5)$
$= 0,4342944819\ldots$

Vega performs this calculation for all prime numbers up to 100.000; all further prime numbers are the result of:
$\log(a \times b) = \log(a) + \log(b)$.

XXXII.   Jurij Vega (III):

In 1793 Jurij Vega attracted the mathematical world's attention one more time by determining 136 decimal places of π. Only thirty years later this record was broken.
Also in this case Vega made use of the series expansion of a function:

$$\arctan(x) = x - (x^3/3) + (x^5/5) - (x^7/7) + \ldots - \ldots$$

Furthermore Vega made use of the addition theorem of the tangent:
$\tan(a+b) = (\tan(a) + \tan(b))/(1 - \tan(a)\tan(b))$.
Form this theorem he deduced the relation:

$$\Pi = 20\arctan(1/7) - 8\arctan(3/79)$$

XXXIII.   The Möbius Function:

August Möbius presented his function in 1831. He claimed that the distribution of all whole numbers could be imagined as all those numbers belonging to three different big boxes.

First box:

Box (0):

In this box belong all manifolds of square numbers, except 1

→ {4, 8, 9, 12, 18, ...}

Second box:

Box (-1):

The second box contains all numbers which are decomposable into an uneven number of prime numbers.

→ Thus: 30 = 5 x 2 x 3 belongs into this group

Third box:

Box (+1):

The third box contains all numbers which are decomposable into an even number of prime numbers.

→ Thus: 6 = 2 x 3 belongs into this group.

Möbius now found out that the probability of a number belonging into box (-1) and box (+1) is equally: $3/\pi^2$

Möbius' function is nowadays particularly applied in the area of Particle Physics.

XXXIV: Srinivasa Ramanujan (1887-1920):

Srinivasa Ramanujan was doubtlessly one of the greatest mathematicians of all times.

Unfortunately he already died at a very early age.

In 1914 the Cambridge mathematician Harold Hardy (1977-1947) invited Ramanujan to come to England in order to work with him.

Ramanujan's work was so successful that he was accepted as a member of Trinity College and Royal Society in 1917.

In 1920 Ramanujan who suffered from tuberculosis was taken to hospital in London where he finally died.

During his time in hospital Hardy visited his friend and colleague as often as he could.

Once when he came to hospital he told Ramanujan that this time he had taken a taxi with the number 1729, a truly boring and uninspiring number.

Ramanujan looked at Hardy and told him that he found the number 1729 extremely interesting.

1729 is the first whole number which can be written as the sum of two cubic numbers:

$1729 = 1^3 + 12^3$ and also

$1729 = 9^3 + 10^3$.

Furthermore 1729 can be divided by its digit sum 19:

$1729 = 19 \times 91$

Ramanujan concluded that 1729 is not such a boring number after all.

XXXV: The Italian mathematician Giuseppe Peano:

Peano claimed that arithmetic should be based on a solid foundation of axioms. He started off by describing a first rather simple set of three axioms:

Cipher is a number.

The successor of a number is itself a number.

There are no two numbers with an identical successor.

Kurt Gödel on arithmetic and logic:

In each formal system S, in which a certain degree of elementary arithmetic is possible, there is at least one undecidable theorem - a theorem, which cannot be proved in S and the negation of which can also not be proved in S.

No formal system S, in which a certain degree of elementary arithmetic is possible, can be consistently and without contradiction proved within itself.

XXXVI: The Catalan Conjecture:

Eugène Charles Catalan combined the squares of integral numbers with the sequences of cubic numbers:

Squares: $\{4, 9, 16, 25, \ldots\}$
Cubic numbers: $\{8, 27, 64, 125, \ldots\}$

$\rightarrow$ The combination of the two sequences leads to the following list:
$\{4, 8, 9, 16, 25, 27, 36, \ldots\}$

In 1844 Catalan formulated the conjecture that 8 and 9 are the only squares/cubes following each other in immediate succession.

Catalan was sure that for the equation $x^p - y^q = 1$; with x, y, p and q > 1 there is only one solution:
$3^2 - 2^3 = 1$

Only in the year 2002 Preda Mihăilesu from the University in Paderborn (Germany) could prove that the Catalan Conjecture is valid.

XXXVII.   Cantor-Conjecture:

If you have a set of natural numbers (1, 2, 3, 4, 5, ...), this set consists of several subsets, such the set of all even numbers (2, 4, 6, 8, 10, ...), the set of all uneven numbers (1, 3, 5, 7, 9, ...) and the set of prime numbers (2, 3, 5, 7, 11, ...). Cantor now found out that an even, an uneven and also a prime number can be allocated to each natural number. This lead Cantor to the assumption that there are as many even, uneven, natural or prime numbers as there are numbers altogether.

XXXVIII.   The Conway Sequence:

If you look at the sequence 1, 11, 21, 1211, 111221 etc. you can see that even if you start the sequence with 1 being the initial number, the sequence increases rapidly. The 16th item of the sequence is already relatively long:

1321132132211331121321133112111312211213211312111322211231131122211311112311 33211121321132221131211321211

John Horton Conway now found out that in this sequence the 1 occurs dominantly, 2 and 3 appear less often and there is no item in this sequence > 3. Conway now declared that if L$\vee$n is the length of the nth item in this sequence, L$\vee$n increases proportionally to the nth power of the Conway constant (1,303577269034269391257099112152551890730702504 6594 ...)

This conjecture is valid for all Conway Sequences starting with any number as initial item except number 22.

XXXIX.   The Brun Constant:

Prime numbers are often immediately succeeding uneven natural numbers like 3 and 5 or 17 and 19. In other words prime numbers often appear as prime number twins. There seems to be an infinite number of prime number twins, the biggest ones having more than 58.000 digits.

Whereas the prime numbers seem to converge at infinity, Brun, in 1919, found out that the reciprocals of succeeding prime number twins converge at a certain limit value.

Thus B = {1/3 + 1/5 + 1/7 + ...} ≈ 1,902160
This value B (1, 902160) is from 1919 onwards called the Brun constant.

XL.   The Birthday Paradox:

In 1939 the Austrian mathematician Richard von Mises raised the question of how many people have to be within a room, so that the probability of two or more people having their birthday on the same day is more than 50%.

He proved that with 365 days per year only 23 persons are needed to raise the probability of two or more of them having their birthday on the same day to 50%. With 57 persons the probability is already $\approx$ 99%.

For a probability of 100% however 366 persons have to be within the room. Von Mises developed the following formula for calculating the probability with which $\geq$ celebrate their birthday on the same day:

$1 - [365! / [365^n (365-n)!]$

XLI. Paul Stäckel and the Twin Primes:

We know that there are no immediately consecutive numbers except the pair $\{2, 3\}$.

This is obvious since 2 is the only even prime number.

Two consecutive uneven numbers like the pairs $\{3, 5\}$, $\{5, 7\}$, $\{11, 13\}$, $\{17, 19\}$ are called twin primes.

The term was first used by Paul Stäckel (1892-1919).

Twin primes are of the kind $[p, p + 2]$.

Further twin primes are:

$\{29, 31\}$, $\{41, 43\}$, $\{59, 61\}$, $\{71, 73\}$, $\{101, 103\}$,
$\{107, 109\}$, $\{137, 139\}$, $\{149, 151\}$, $\{179, 181\}$, $\{191, 193\}$,
$\{197, 199\}$, $\{227, 229\}$, $\{239, 241\}$, $\{281, 283\}$ ...

There seems to be an infinite number of twin primes.

The largest pair of twin primes found so far is:

$(33.218.925) \times (2^169.690) - 1$

and

$(33.218.925) \times (2^169.690) + 1$.

These 51.090 digit numbers were detected in the year 2002.

XLII. The ABC-Conjecture:

Joseph Oesterlé and David Masser suggested the ABC-Conjecture in 1985.
They defined numbers as square and square-free numbers respectively.

Whereas 13 is square-free, 9 is a square number (3^2).
The square-free part of a number is described as sqp (n).

Since the prime factors of 15 are 5 and 3, 15 is also square-free.
With n = 8 all prime factors included are 2; thus sqp (8) = 2

With n = 18 the prime factors are 3 (2 x) and 2; thus sqp (18) = 6

If the prime numbers A = 3, B = 7 with their sum C = 10 have no factors in common and are square-free numbers, their product is also square-free: 210.

The ABC-Conjecture now claims that [sqp (ABC)] ^n/C possesses a positive lower limit bigger than 1, if n is a real number > 1.

XLIII.   The Andrica Conjecture:

The smallest examples of prime numbers are 2, 3, 5, 7, 11, 13, 17, 19, 23, 29, 31, and 37.

Leonhard Euler was one of the first mathematicians who became interested in the arrangement of prime numbers and also in the gaps between prime numbers. In 1985 Dorin Andrica found out that there is a gap consisting of 879 non-prime numbers following the prime number 277.900.416.100.927.

In the 2009 a new non-prime numbers gap was discovered consisting of altogether 337.446 non-prime numbers.

In his conjecture Andrica made the following approach: If $\sqrt{(p\vee (n+1)} - \sqrt{p\vee n} < 1$ is given with $p\vee n$ being the n-th prime number, the gaps between prime numbers could be calculated as follows:

$g\vee n < 2 \sqrt{(p\vee n)} + 1$ with $g\vee n$ being the n-th prime number gap and $g\vee n = p\vee (n + 1) - p\vee n$.

In the year 2008 Andrica's Conjecture could be validated for:

$n < 1,3002 \times 10^{16}$